Rocks and Plants

A Pocket Field Guide to the Geology and Botany of the St. George Basin

by Eric Hansen

Thanks

[signature]

348 North Lodge Road
Central, UT 84722

Copywright © 1997
by Eric Hansen

ISBN 0-9662991-1-6

To Rocky and Sunny

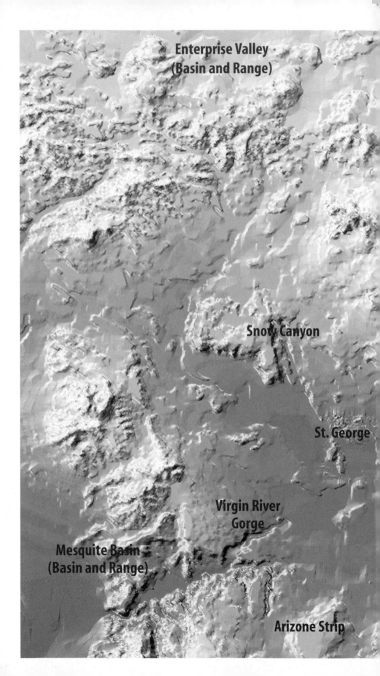

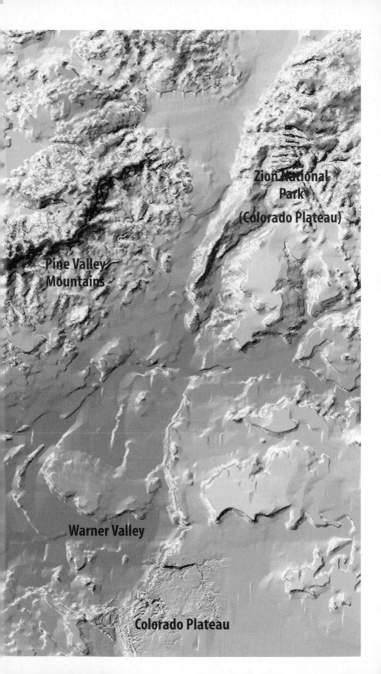

ACKNOWLEDGMENTS

This little guide went through several drafts, some too long and complicated, some too simplified. I persuaded nearly everyone I know to read it, usually against their will. Mom agonized over my grammer. Randy reformatted the entire thing and threw out almost every picture. Lucky for me, Bryce let me pilfer his photo archive. Charisse eliminated two whole chapters. Lonnie made me rewrite lots of the geological material. Marc went so far as to put the guide to the test with the hiking staff at Green Valley Spa. And there were some - Lara, Dave, Steph, Amanda, Thad, Alan, Martha, Cheryl, Todd, Karen, Jay, and others - who just thought it was really cool. The following sources provided a valuable source of information and deserve acknowledgment:

Barnes, F. A. (*Canyon Country Geology,* Wasatch Publishers, 1978),

Brown, David E. (*Biotic Communities: Southwestern United States and Northwestern United States and Northwestern New Mexico.* University of Utah Press, SLC, 1994).

Bugden, Miriam (*The Geology of Snow Canyon.* Department of natural Resources, Utah Geological Survy Public Information Series 13, 1992).

Budgen, Miriam (*Geologic Resources of Washington County, Utah.* Department of Natural Resources, Utah Geological Survey Public Information Series 20, 1993).

Chronic, Halka (*Roadside Geology of Utah.* Mountain Press Publishing Company, Missoula Montana, 1990).

Crawford, J.L.. (*Zion National Park: Towers of Stone.* Zion Natural Histroy Association, Springdale, Utah, 1994).

Eardley, A. J. and James W. Schaack (*Zion: The Story Behind the Scenery.* KC Publications, Inc., Las Vegas, 1996).

Harrington, H.D. (*Edible Native Plants of the Rocky Mountains*. University of New Mexico Press, Albuquerque, 1993).

MacMahan, James, et al. (*Deserts: Autobon Society Nature Guides*. Alfred A. Knopf, Inc, New York, New York, 1985).

Nations, Dale and Edmund Stump (*Geology of Arizona.* Kendall/Hunt Publishing Company, Dubuque, Iowa, 1981).

Procter, Paul D. (*Geology of the Silver Reef Mining District.* Utah Geological and Mineral Survey, Bulletin No. 44).

Stokes, William Lee (*Geology of Utah.* Utah Museum of Natural History, University of Utah, SLC, 1988).

Welsh, S.H. et al. (*A Utah Flora.* Great Basin Naturalist Memoir No. 9. Brigham Young University Press, Provo, UT, 1987)

CONTENTS

Snow Canyon State Park

Coelophysis footprint in Moenave Mudstone,
Johnson Farm, St. George, Utah

INTRODUCTION

The Dixie Basin is situated in a transition zone where two major physiographic provinces meet (Basin and Range and the Colorado Plateau). The Beaver Dam Mountains to the west and the Hurricane Cliffs to the east form the local physiographic boundaries. So, what's in between? St. George, Snow Canyon, the Pine Valley Mountains, ancient floodplains and terraces, synclines and anticlines, faults and folds, laccoliths, cinder cones and extinct volcanoes, eroded canyons and mountains of metamorphic, igneous, and sedimentary origins: a big geological conundrum, in other words. It challenges geologists and attracts outdoor enthusiasts and naturalists from all over the world.

In the midst of this transition in land forms, three of North America's deserts collide. These include the Mojave Desert, the Great Basin Desert, and the so-called Colorado Plateau Semi-Desert. Each of these deserts are made up of many different biotic communities (groups of plants and animals that generally occur together). Our area's proximity to the three deserts, and the drastic changes in elevation associated with the rugged topography, give us a smattering of species representative of biotic communities from all three deserts. In addition changes in temperature and elevation occur so quickly, and within such short distances, that even non-desert species from the surrounding highlands sneak down the mountains and cohabitate with their long-lost desert relatives. The resulting biotic diversity is remarkable and nearly overwhelming to ecologists, who like to keep their flora and fauna organized into regional categories.

In summary, the area's transitional location between physiographic provinces and at the intersection of three of North America's deserts results in biotic and geological goulash. This booklet is designed to help simplify some of the confusing, but intriguing natural relationships and assist visitors and natives alike in the exploration and enjoyment of this unique and beautiful area. It's divided into two parts: Part I discusses the local development of the region's most prominent geological formations; Part II reviews the local desert biotic communities (desert flora and fauna) and provides a plant identification key that includes identification characteristics of most of the region's woody desert plants. Also included are two appendices (Appendix A and Appendix B). Appendix A provides descriptive information on identification characteristics and possible uses of the plants referred to in Part II. Appendix B includes a brief, simplified introduction to the geology of the different types of rocks and minerals found in our area.

Springtime in the Mojave Desert, South of Green Valley.

Red and white hues of Navajo Sandstone, Snow Canyon.

Eroded *"hoodoos"* of Navajo Sandstone, near Leeds.

PART 1: ROCKS

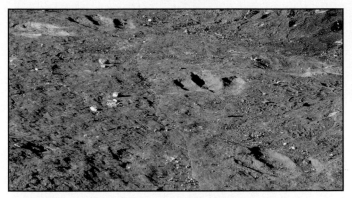

Prosaurapod foot prints in Moenave mudstone, Warner Valley.

THE HISTORY

Most of the rock formations exposed in our area were created between 250 to 180 million years ago. During the beginning of this time period, what is now the southwestern United States was located near the equator and made up a small portion of Pangea, a "super-continent" that was made up of all the known continents of the world. Elevation was near sea-level, a low spot where gravity discharged sediment and water from surrounding highlands (e.g., the Ancestral Rockies to the east, the Mesocordilleran and Mogollon Highlands to the west and south, respectively).

During this time period, the accumulation of sediments in our area was extensive. Sediments were transported here via streams, rivers, oceans, and wind currents, often over thousands of miles, and deposited on river terraces, ocean flats, flood plains, stream banks, dry basins, and at the bottom of large bodies of fresh water. Consider the quantity of newly deposited silt and mud transported from Colorado Plateau to the bottom of large man-made reservoirs such as Lake Powell or Lake Mead over the course of the last 50 years or so. Recent measurements indicate that over a hundred feet have accumulated in some areas. Now imagine rivers the size of the Amazon, lakes larger than Lake Michigan and vast inland seas all carrying immense quantities of sand and silt into what is now the Southwest over tens of millions of years.

You may begin to fathom the magnitude of the geological process. Geologists estimate that up to twelve thousand feet of earth was transported by wind and water into the region. The sheer weight of the newly deposited materials created subsidence in the earth's crust originating at the mantle. This adjustment kept ground surface elevations at or below sea level, pushing older materials deep into the earth, allowing gravity, water, and wind to maintain a vertical accumulation of land mass at the surface.

These sediments were rich in iron and manganese, minerals responsible for coloring the mesas and canyons of the region. Iron, for example, in its unoxidized state, creates hues of green and blue especially conspicuous in shale and mudstone beds. Iron is prevented from oxidizing under anaerobic conditions such as swampy environments where decay of abundant organic materials prevents oxidation. Oxidized iron minerals, such as hematite, create reds, pinks, oranges and yellows. Manganese creates darker surfaces that appear lavender, purple or shiny black and brown. This is contrasted by the whites and tans of calcium carbonate minerals in limestone and the blacks and grays of volcanic materials.

Under enormous compaction pressure from suerimposed land mass, the buried materials eventually lithified into layers of sedimentary rock -- geological formations made up of sandstone, shale, siltstone, and limestone. A more detailed explanation of sedimentary rock and the lithification process is discussed on page 72. A stratigraphic cross-section showing the relative vertical arrangement or superimposition of some of the prominent geological formations in our area is shown on the opposite page. The age of the formations presented here are approximations and will vary according to different reseachers. A simplified plan-map of where these formations are exposed in our area is presented on pages 8-9.

The sequence of events that created the canyon country landscape during the Triassic and early Jurassic Periods has been studied extensively by geologists, paleontologists, and paleogeographers. While the details are constantly being debated and updated, a brief, generalized synthesis of the events that occurred in our area during this period follows.

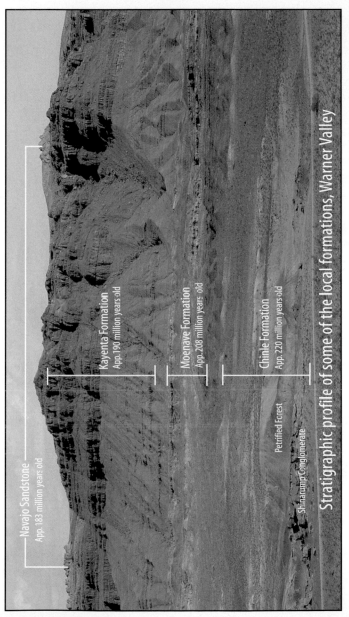

Navajo Sandstone
App. 183 million years old

Kayenta Formation
App.190 million years old

Moenave Formation
App. 208 million years old

Chinle Formation
App. 220 million years old

Petrified Forest

Shinarump Conglomerate

Stratigraphic profile of some of the local formations, Warner Valley

ROCKS

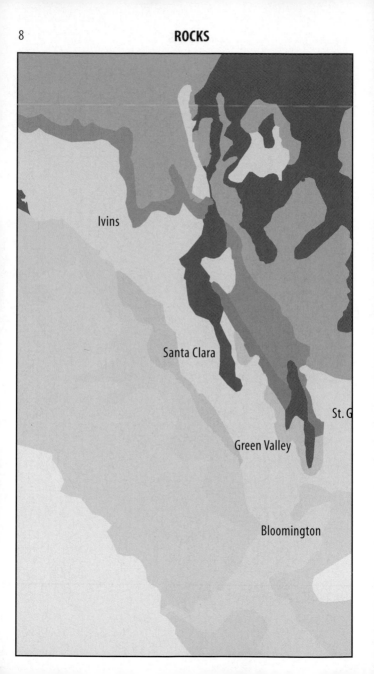

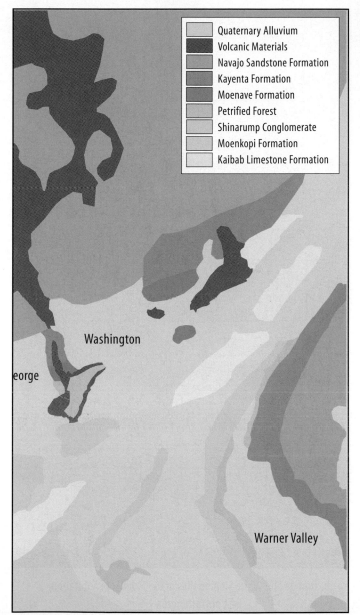

Orange and White Sloping Hills of the Moenkopi Formation,
East of Washington Fields.

THE MOENKOPI FORMATION (245-220 MYA)

The shoreline of the Pacific Ocean was once closer than it is today. During the early Triassic, it occasionally intruded into central Utah through a broad corridor that ran north-northeast through the state. The waters would retreat periodically and a tropical climate would prevail, nurturing a biological wonderland with a diverse assemblage of plants and animals. Sediments transported into our area as a result of the tropical and marine environments of the early Triassic are collectively referred to as the Moenkopi Formation.

In our area the Moenkopi is the first geological formation representing the age of the dinosaurs. It sits atop Kaibab limestone, the cap rock of the Grand Canyon and the last formation laid down during the age of amphibians. Geologists suggest that the source of most of the Moenkopi Formation materials originated from the "Ancestral Rockies," an ancient mountain range that encompassed portions of the extant range, including portions of eastern Utah and western Colorado. This formation is about 1,750 to 2,320 feet thick and consists of numerous narrow layers of sandstone, shale, gypsum and limestone. It varies in color from red, brown, pink, gray, and white, but in our immediate area appears mostly light orange to reddish. There are numerous different members of this formation, each with its own array of minerals. Consequently, erosion effects each member uniquely. The result is a bizzare landscape of miniature cliffs, small sloping mounds, massive, sheer mesas and montains. The lowermost layers contain oil depoits. Oil wells in the town of Virgin, UT have tapped into this energy resource with some success.

Fossils encountered in this geological formation include numerous lizard-like reptiles and lots of amphibians, as well as an abundance of marine crustaceans, such as snail-like creatures called Meekoceras, Tirolites, and Columbites; and organisms similar in appearance to sand dollars called Unionites, Lingulas, and Clarias. It was a transitional period when nature was busy experimenting with new, land-loving, scaley, cold-blooded life forms that would become the dominant life form on the planet for the next 200 million years. Good views of this formation can be attained nearly anywhere south of St. George and adjacent to I-15 between the Virgin River Gorge and Bloomington or on the Arizona Strip.

Stratigraphic profile of the Moenkopi Formation, Bloomington Hills. Shinarump Conglomerate forms the caprock.

Shinarump Conglomerate outcrop (foreground) and sloping, Petrified
Forest ash bed (center) of the Chinle Formation, Warner Valley.

THE CHINLE FORMATION (220-208 MYA)

The Middle Triassic (230 – 220 million years ago) was a period of erosion
and mountain building. Very little geological evidence from this period
remains in the Southwest. Erosion not only washed away whatever Middle
Triassic deposits may have existed, but removed a significant portion of the
Moenkopi Formation below it as well. During this period, tectonic activity to
the south, west, and north created highlands in Arizona (Mogollon
Highlands) and northern and western Utah (Mesocordilleran Highlands –
Sevier Range). The eventual and inevitable result was the formation of a
large, land-locked basin that encompassed most of the southern and east-
ern portions of the state.

By the end of the middle Triassic, the highlands to the south and west had created a barrier that obstructed ocean waters from entering the region. The Mogollon highlands to the south, and the Uncompagre Highlands to the west and north, were being uplifted. Vast quantities of sand and gravel were shed of these emerging highlands into southwestern Utah and northern Arizona.

> Some geologists (Stokes, *Geology of Utah,* Utah Museum of Natural History, 1988 -- and Barnes, *Canyon Country Geology,* Wasatch Publishers, Inc. 1978) suggest that the source material of the Chinle and subsequent Triassic formations originated from the remnants of the Uncompagre Uplift and the Ancestral Rockies to the east, pointing to evidence that the streams flowed from the east. Others (Nations and Stump, *Geology of Arizona,* Kendall/Hunt Publishing Company 1981) argue that since that the sand and gravels of the Chinle Formation came from paleozoic rock units known to crop out in Central Arizona and that since these sediments are coarser in the south and finer as distance increases to the north, that the Mogollon Highlands in Central Arizona must have been the source.

These relatively coarse sediments, known as the Shinarump Conglomerate, were spread uniformly over the eroded, early Triassic Moenkopi formation below it. As erosion eventually decreased the elevation and gradient of the Mogollon Highlands, finer sediments were laid down by streams and rivers over the Shinarump Conglomerate forming multicolored shales and sandstones. Large lakes trapped in the basin created limestone and gypsum deposits. Tectonic activity, responsible for the uplifts, also resulted in volcanic eruptions, which spread immense quantities of ash over the area. Collectively, the sandstones, shales, lime stones, and ash beds deposited during this period are referred to as the Petrified Forest member of the Chinle Formation.

Fossils of the oldest known dinosaur, Coelophysus, were encountered in the Chinle Formation in surprising numbers. They were only slightly larger than turkeys, with hollow bones and razor sharp teeth. They hunted in packs and were one of the most formidable predators of their time. Also discovered in the Chinle Formation are fossilized remains of North American crocodiles, phytosaurs, and other dinosaurs. Locally, sandstone outcrops of the Chinle

Formation in Silver Reef have been mined for uranium, copper, and silver. During the cold war, uranium mines opened in many parts of southern Utah, especially southeastern Utah and near Emery, where uranium was mined from the Chinle Formation. The Silver Reef Mining District in Leeds remains the only known source of commercial bodies of silver ore recovered from sandstone in the United States.

Good exposures of the Shinarump Conglomerate are found west of Green Valley, Santa Clara, and Ivins. It forms the caprock of the sloping hills to the west. Shinarump cliffs are also conspicuous south of St. George and atop the Washington Dome, east of Washington. The Petrified Forest Member occurs sporadically throughout our area and is often scarred with off-road-vehicle tracks. It usually appears as sloping mounds or hills and looks purplish from a distance. It also contains "Blue Clay". The clay is primarily decomposed volcanic ash called bentonite. The bentonite clay expands when wet and contracts when dry causing cracking of concrete foundations. By the end of the middle Triassic, the highlands to the south and west had created a barrier that obstructed ocean waters from entering the region. The Mogollon highlands to the south, and the Uncompagre Highlands to the west and north, were being uplifted. Vast quantities of sand and gravel were shed of these emerging highlands into southwestern Utah and northern Arizona.

Coarse sandstone and conglomerate rock (right) of the
Shinarump Conglomerate, West of Bloomington.

Bright purple, sloping mounds of the Petrified Forest
Member of the Chinle Formation, Warner Valley.

Erosion creates a statue in the Shinarump
Conglomerate, South of Green Valley.

THE MOENAVE FORMATION (208-190 MYA)

Millions of years of runoff from the Mogollon Highlands eventually created numerous bodies of water and slow moving streams and rivers in the basin. Lakes, swamps, ponded drainages, and ponds covered large portions of southern Utah and Northern Arizona. These bodies of water slowly accumulated sediments carried by air currents, slow-moving streams and rivers. Sediment particle sizes were sorted in the lakes. Sand and gravels, the heaviest particles, were amassed near shoreline, where their greater relative weight allowed them to sink in lake currents and waves, while the smaller particles remained suspended until reaching relatively calmer waters further from shoreline. The result was the creation of numerous interbedded layers of sandstones, shales, siltstones, and mudstones collectively referred to as the Dinosaur Canyon Member of the Moenave Formation. Locally, Dinosaur Canyon deposits reach a maximum thickness of 570 feet.

Open exposures of this member appear to consist of giant bathtub rings (narrow layers of shale and mudstone sandwiched between relatively larger sandstone layers). The sandstones are generally reddish brown to lavender, while the shales and mudstones often appear greenish to gray. Erosion affects each of the sandwiched layers differently creating a landscape of steep slopes with undulating mounds, crumbly ledges, sheer cliffs and overhangs. Faster moving streams and river deposits accumulated during the latter stages of the Moenave Formation. The Springdale Sandstone Member represents this period. It is usually seen as vertical cliffs capping the steep slopes of the Dinosaur Canyon Member below it. The formation varies in color from tan to purplish-pink and measures up to 100-feet thick in some areas.

Fossils of late Triassic life forms abound in the Moenave. Locally, footprints of dinosaurs such as *Coelophysus*, *Dilophosaurus*, and different sauropods have been found in the Dinosaur Canyon member. The numerous dinosaur fossil casts recently discovered at Johnson Farm on Foremaster Road behind

Colorful sloping hills and ledges of the Moenave Formation. Kayenta Formation cliffs emerge in the background, Warner Valley.

the Red Cliffs Mall are a prime example. The formation also contains an abundance of petrified wood. Channels cut into the formation often reveal root systems and an abundance of mineralized organic debris from lake and swamp environments of the period. A relatively complete fossilized gar-like fish called *Semionotus kanabensis* was recovered from the Springdale Sandstone Member of this formation near Springdale, UT.

As distance from the ancient highlands increases, the lake and stream deposits of the Moenave are gradually replaced by aeolian, or wind-deposited sandstones of the Wingate Formation. The Wingate formation is not represented in our immediate area, but covers a large portion of greater Canyon Country. It was deposited in Sahara-like desert environment and consists mostly of desert dunes and dry lake beds. The Moenave merges or inter-fingers with the upper deposits of the Wingate Formation, suggesting that during the latter stages of the Wingate deposition, the two formations were accumulating simultaneously.

This stratigraphic and hence, temporal association may indicate that the Moenave accumulated in a relatively arid climate and that the Moenave lakes and ponds were fed mostly by runoff from the surrounding highlands, which supplied the adjacent lowlands with enough water to maintain an almost tropical or swamp-like landscape. It is also likely that the rising highlands to the west, either an early manifestation of the Sierras, or a closer range referred to as the Mesocordilleran High, had begun to create a formidable rainshadow, not unlike that created by the modern Sierras over the Great Basin.

Exposures of the Moenave are revealed in road cuts carved into the steep, slopes of Black Hill and Formaster Ridge. But the formation is best viewed along the foothills of Sand Mountain in Warner Valley.

Dylophosauras cast in Moenave Mudstone,
Johnson Farm, St. George.

Moenave strata sandwiched between purple Petrified Forest ash and
Kayenta Sandstone ledges, Warner Valley.

Kayenta formation slopes and cliffs, Ivins.

THE KAYENTA FORMATION (190-183 MYA)

Near the end of the Triassic, the climate was becoming increasingly arid. North America had begun to separate from Pangea and move northward. What is now the Southwest was slowly floating into "horse lattitude" (between the 15th and 30th parallels), the lattitude currently occupied by the Sahara Desert and some of the driest regions on earth. During this period the region probably had a climate similar to that of the Senegal, with wet summers and dry winters but with increasing aridity as it was slowly consumed by a vast northern desert. Notwithstanding the increasingly arid conditions, what appears to be a large and prosperous population of dinosaurs left behind an abundance of evidence of their existence in the form of foot prints in the Kayenta Mudstone.

Sluggish streams followed the high energy streams that deposited the Springdale Sandstone. These streams transported and deposited up to 500 feet of iron-rich sand and silt over most of what we call Canyon Country. As the iron in the sediments oxidized, it created red, purple and maroon-colored sand and mudstones. The Kayenta Formation materials often bear fossilized ripple marks, cross beds, and channel scours characteristic stream and river deposits.

Kayenta sandstone is harder than the sandstone above (Navajo Sandstone) or below it (Springdale Sandstone). Unlike the Navajo Sandstone above it, the Kayenta is waterproof. This forces water stored in permeable Navajo sandstone to move horizontally when it comes in contact with the Kayenta, often resulting in desert springs and seeps that support a diversity of plant life. Weeping Rock in Zion National Park is probably the best local example of a seep emerging from the Navajo-Kayenta contact. The cliffs and ledges of Red Hill, located immediately north of St. George are also situated near the point of contact. The presence of seeps or springs in this area were probably influential in the settlement of the area.

One of the best examples of the Kayenta formation is "Red Mountain" in Ivins. The Kayenta formation comprises most of the lower two-thirds of Red Mountain and grades almost imperceptibly into the Navajo Sandstone Formation above it. The lowermost sediments form steep slopes, while the upper portion forms sheer cliffs or caprock.

Kayenta cliffs form a majestic barrier behind Ivins and Kayenta.

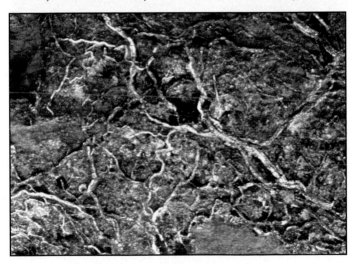

Petrified root system from possible Triassic
swamp in the Kayenta Formation, Warner Valley.

Red and white cliffs of Navajo Sandstone, Snow Canyon.

NAVAJO SANDSTONE (183-170 MYA)

By the end of the Triassic, the semi-arid climate that dominated during formation of the Kayenta Sandstone had given way to extreme aridity. Canyon
Country was swallowed whole in a Sahara-like desert environment.
Climatic conditions had changed to the point where wind, rather than
water, became the chief agent of deposition and for the next 13 million
years, windstorms blew in sand. The result was the eventual accumulation
of one massive sand pile that blanketed most of Utah, and portions of
Arizona, Nevada, and southern California. Locally, the sand pile reached
depths of up to 2000 feet. Such monumental sand heaps are exposed in
profile in the cliff faces of Snow Canyon and such majestic geologic monu-
ments as West Temple, Alter of Sacrifice, Checkerboard Mesa, and the Court

of the Patriarchs at Zion National Park. Navajo Sandstone is not as well indurated as some of the water deposited sediments of earlier and later periods. It's porous enough to be water permeable and serves as a natural reservoir or aquifer of fresh water. Prior to the construction of Quail Creek Reservoir, the city of St. George relied heavily on this giant stone reservoir. Its porous nature also makes Navajo Sandstone friable or crumbly, leaving this formation extremely vulnerable to erosion. Wind and water have have beaten this formation into surreal landscapes of deep, sheer-walled canyons, spires, and domes. Most of the famous geological features of Snow Canyon and Zion National Park are carved from sediments making up this formation.

Landscapes sculpted by erosion into Navajo Sandstone appear almost alien. Bare mounds and sloping hills of solid bedrock colored bright red, pink, orange and white; multitudes of narrow, meandering channels with vertical walls creating elaborate stone labyrinths: spires and hoodoos emerging out of bedrock; isolated ponds that hold water only a few months a year while almost magically supporting large populations of tadpoles and frogs; the occasional gnarled juniper finding footing in a small crack half way up a 300 foot sheer cliff. This type of landscape is usually what is referred to as "slickrock". Wherever it is found it immediately begins to draw the attention of outdoor enthusiasts like hikers and mountain bikers. While it looks friendly enough, the crumbly nature of the sandstone makes it difficult for climbers and hikers to find solid footing or hand-holds, and results in numerous fatal accidents every year.

The lower portion of the Navajo Sandstone generally appears reddish, while the upper portion appears white. This is due to the presence of hematite (an iron oxide or "rust" mineral) in the former, and its absence in the latter. Some geologists suggest that, over time, gravity and water leached the iron minerals from the upper portions of the formation. Others suggest that the original sand source simply ran out of iron minerals while it was being deposited. The latest theory postulates that hydrogen sulfide gas, produced from relatively recent volcanic activity, "sulphidized" or reduced the oxidized iron minerals in the upper portions of the formation, hence, neutralizing or whitening the reddish hematite.

Sometimes, concentrations of Iron and manganese oxide form small spheres or concretions locally referred to as "Moki Marbles." The metallic concretions are heavier than the surrounding sandstone and make a metallic "clank" when they collide. They often erode out of the sandstone and accumulate in large quantities in low spots. When scattered loosely on trails of naked bedrock, they pose an interesting obstacle for hikers.

Moki Marbles eroding out of the Petrified Dunes in Snow Canyon.

Often, the sandstone turns dark maroonish or shiny black and appears polished. This phenomenon is usually referred to as "desert varnish." Desert varnish results from a residue or film of manganese and/or iron left on the sandstone by rain or groundwater. Recent evidence suggests that bacteria on the sandstone are instrumental in extracting and concentrating the manganese residue on the cliff walls.

Checkerboard patterns often appear on Navajo Sandstone and are generally the result of erosion exposing the original dune sand beds accumulated during periods of prolonged wind shifts (the horizontal lines), and accelerated erosion along steeply dipping joints in the frlable sandstone (the vertical lines).

Desert Varnish on Navajo Sandstone in Snow Canyon.

Checkerboarding revealed in the "Vortex", near Gunlock Reservoir.

Cinder Cone near Diamond Valley,
volcanic feature associated with the most recent eruption.

THE REST OF THE STORY

For the most part, the geological record in the St Ge
with the Moenkopi Formation and ends with Navajo Sa
the story is 170 million years long and begins with th
down on top of the Navajo Sandstone, mostly at higher
of the St. George Basin. However, several events during th
related to the formation of the modern topography of
overview of these follows:

The Sevier Orogeny

The separation of the North American continent from
earnest about 150 million years ago. North America be
westward, forcing its way against and over the ocean
plate. The pressure created by the movement of the co
extreme crustal deformation along the western margin
from southern Nevada to Alaska. The land literally buckl
sure. The continental crust folded upon itself and was
sixty miles from overthrust during the event, which is th
ed for a period of 25 million years.

This crustal deformation, referred to as the Sevier Thrus
mountains to the north and west of canyon country an
sification of volcanic activity in the region. The mounta
impenetrable barrier to Pacific Ocean waters invading 1
furnished additional raw materials that contributed s
augmentation of Cretaceous and Tertiary sedimentary 1

Remains of the effects of the Sevier Orogeny include
that created the valley running north from Kanab (High
Junction that separates the Markagunt Plateau (Bria
Breaks) and the Pansagunt Plateau (Bryce), and the Ka

The Virgin Anticline, crustal deformation
caused by the Laramide Orogeny.

The Larimide Orogeny

About 80 million years ago, the North American continent once again began to move in a different direction and at a greater velocity than it had done before. The event initiated what geologists call the Laramide Orogeny. As the continent rode westward over the Pacific oceanic plate, the western half of the continent experienced an interval of upheavals, earthquakes, volcanoes and crustal deformation that would last for more than 30 million years. During this chaotic period, the California coastal mountain range and the Rocky Mountains were born. The Sierras attained new heights, and the entire western half of the continent, from the pacific coast to the rocky mountains, was uplifted, exposing sediments deposited tens of millions of years earlier. Isolated areas in Canyon Country experienced larger uplifts, such as the San Raphael Swell, the Monument Upwharp, the Uncompagre Uplift and the Circle Cliffs.

Locally, the Laramide Orogeny produced crustal deformations such as the Virgin Anticline, a comb-like ridge visible from I-15 between Hurricane and St. George, Square Top, north of Gunlock, and the Beaver Dam Mountains south and east of town.

Square-top, a crustal deformation resulting
from the Laramide Orogeny.

The Basin Range Extentional

About 18 million years ago, pressure between the Pacific and North American plates subsided due to the formation of a transform fault on the west coast. Known as the San Andreas Fault, it relieved pressure between the two plates by allowing the pacific plate to slip northeastward in relation to the North American plate. The Gulf of California and the Baja California Peninsula formed as a result of this slippage.

With the subsidence of tectonic pressures against the continental crust came the beginning of an event called the Basin and Range Extensional. The land west of the Wasatch Mountains and east of the Sierras, from northern Mexico to Oregon, began to stretch and actually expand westward. Geologists estimate that this area gained more than 50 miles of new territory during the expansion. The continuous westward movement and stretching of the crust resulted in earthquakes and the adjustment of continental crust along several fault lines.

One large fault line that currently follows the western margin of the Colorado Plateau from nothern Utah to Central Arizona, locally referred to as the Hurricane Fault, opened and released the tension of the extension on the land east of the fault. Freed from the gravitational pull of the land west of the fault, the land east rose in response with a concommitant subsidence or downward isostatic adjustment in the land to the west. In some areas, vertical displacement measures several thousand feet.

The stretching of continental crust also caused blocks of continental crust that had previously faulted, to tilt or sink relative to surrounding blocks, creating a series of north to south trending ranges separated by large, arid basins (i.e., Basin and Range Province). The unique structural setting of the Basin and Range along with contemporaneous volcanic and associated hydrothermal activity resulted in one of the world's significant metalliferous provinces.

The Hurricane Cliffs, the physiographic
boundary of the Colorado Plateau.

Faulting: note the amount of vetical displacement between
the colored strata on the right (Washington Fields)
and upper left (The Hurricane Cliffs).

The Pine Valley Mountain Laccolith, a "volcanic plug".

Volcanic Activity

Around 20 million years ago, and perhaps, partly related to the events surrounding the Basin and Range Extentional, the Pine Valley Mountains were born. Magma forced its way through the earth's crust and traveled quickly toward the surface. The molten rock broke through ancient layers thousands of feet thick, but was ultimately prevented from surfacing when it came into contact with the pink limestone of the Claron Formation, the same formation exposed in the sculptured surrealistic landscapes of Bryce Canyon and Cedar Breaks. The Claron Formation was created at the bottom of Lake Flagstaff, a fresh-water sea that once covered a large protion of southwestern and easttern Utah between 50 and 35 million years ago. It is the youngest of the major formations exposed in our area.

The rising magma could not penetrate the Claron Limestone and instead spread laterally within it, forcing the overlying strata to arch, creating an enormous dome with an igneous granite heart and a surprisingly plastic limestone exterior. As the limestone above the granite eventually eroded away, the granite interior was exposed. This "volcanic plug," or "laccolith," is highly visible from anywhere in the St. George Basin and has a remaining thickness of over 3,000 feet and covers over 70 square miles of land, the second largest designated wilderness area in Utah.

The lava flows in and around Snow Canyon appear to represent three periods of volcanic activity. The first took place around three million years ago. Lava filled river and stream channels and blanketed portions of the area around Snow Canyon with a protective cap-rock of basalt. The intruding volcanic rock altered existing water courses, forcing erosive activity to the margins of basalt flows, eventually carving canyons into the adjacent unprotected, sedimentary rock. Subsequent lava flows repeated the process. The most recent occurred within the last three thousand years and is associated with the cinder cones at the northern end of Snow Canyon near Diamond Valley. This last eruption covered much of the base of Snow Canyon and the surrounding lowlands with basalt. The youngest volcanic materials are therefore topographically lower than antecedent flows. The result is what geologists refer to as inverted topography.

Remnants of the first flow can be seen forming the terrace above Highway 18 between St. George and the northern entrance of the park. The highway itself was built on volcanic remains of the second flow. The third is the most highly visible and is especially conspicuous near its source, around the Diamond Valley Cinder Cones and in topographic lows in Snow Canyon, Ivins, and Santa Clara.

Black Hill, erosion-resistant basalt from a three
million-year-old lava flow caps this mesa, St. George.

A Joshua tree growing in a juniper woodland, a rare association representing a transition zone between deserts.

PART II. THE PLANTS

BIOTIC COMMUNITIES

The St. George Basin lies at the junction of three deserts: the Mojave Desert, the Great Basin Desert, and the so-called Colorado Plateau Semidesert. Each desert is defined by groups of plant and animal species (i.e. biotic communities) adapted to unique seasonal precipitation and temperature regimes. While many individual species cross desert boundaries, their association with other species occurring in the near vicinity is generally indicative of the desert community represented.

Mojave Desert community with pale cholla, creosotebush,
buckwheatbrush, bursage, ratany, and snakeweed.

The Mojave Desert

The Mojave Desert enters our area through the Virgin River Gorge, from
southern Nevada. It's the driest and the smallest of North America's true
deserts. It includes a large portion of the southern third of inland California,
the southern tip of Nevada, portions of northwestern Arizona, and the
southwesternmost corner of Utah. Generally, precipitation ranges from

between 2.6 to 7.6 inches per year, with the majority of rainfall occurring in the winter. Annual precipitation in Saint George is higher than in most areas of the Mojave Desert. Extreme temperatures range from between 19 degrees In the winter to around 115 degrees in the summer. Mojave species thrive in the St. George Basin, but cannot survive the colder climates that prevail around it.

Mojave Desert plant communities in our area are usually dominated by one or more of the following species: creosotebush *(larrea tridentada)*, black-brush *(Coleogyne ramosissima)*, bur-sage (Ambrosia dumosa), sand sage-brush *(Artimesia filifolia)*, and shadscale *(Atriplex confertifolia)*. Also abundant and conspicuous, though not usually dominant are distintive Mojave residents such as the Joshua Tree *(Yucca brevifolia)*, Anderson thornbush *(Lycium andersonii)*, pale cholla *(Opuntia echinocarpa)*, barrel cactus *(Ferrocactus acanthodes)*, honey mesquite *(Prosopis juliflora)*, catclaw *(Acacia gregii)*, ratany *(Krameria parvifolia)*, narrow-leaved yucca *(yucca angustissima)* indigobush *(psorothamnus fremontii)*, Mormon tea *(Ephedra sp.)*, buckwheat bush *(Eriogonum fasciculatum)* and prickly pears such as beavertail *(Opuntia basilaris), and* Mojave prickly pear *(Opuntia erinacea)*.

Also representative of the Mohave Desert in our area are reptiles such as the desert tortoise *(Gopherus agassizi)*, the sidewinder rattlesnake *(Crotalus cerastes)*, the Mojave green rattlesnake *(Crotalus scutalatus)*, the gila monster *(Heloderma suspectum)*, and the chuckawalla *(Sauromalus jobesus)*.

　　　The Joshua tree has come to symbolize the Mojave Desert. While not the most abundant or widespread species, it is certainly the most conspicuous, growing up to 50 ft tall with giant pointy arms. Early Mormon settlers claimed that the arms pointed the way through the desert to the promised land, like the Biblical Joshua. The geographic limits of the Joshua tree appear to outline the boundary of the Mojave Desert. While the Joshua Tree sometimes occurs slightly outside Mojave Desert boundaries, it prefers this desert almost exlusively.

Joshua tree and pale cholla, Mojave Desert species.

Mesquite, a Mojave Desert species.

Big sagebrush, a Great Basin Desert dominant.

The Great Basin Desert

The Great Basin Desert is the largest of North America's deserts. It encompasses most of Nevada, and portions of Idaho, Oregon, California and Utah. It enters the St. George Basin from the North, but cannot endure the arid climates to the south and west. Rainfall averages between 5.7 to 11.5 inches per year. During most of December, January, and February, average daily temperatures often remain below freezing. Nighttime temperatures during all but the warmest months of the year can dip below the freezing point. Summer temperatures often exceed ninety degrees.

Dominant plant species of Great Basin Desert biotic communities in our area include one or more of the following: Big sagebrush *(Artemisia tridentata)*, rabbitbrush *(Chrysothamnus nauseosus)*, winter fat *(Ceratoites lanata)*, greasewood *(Sarcobatus vermiculatus)*, lambs wool *(Kochia lanata)*, four-wing saltbush *(Atriplex canescens)*, shadscale *(Atriplex confertifolia)*, and blackbrush *(Coleogyne ramosissima)*. Other less dominant, though characteristic Great Basin Desert species occurring in our area include: banana yucca *(Yucca baccata)*, squawbush *(Rhus trilobata)*, cliffrose *(Cowania mexicana)*, Utah juniper *(Juniperus utahensis)*, century plant *(Agave utahensis)*, snakeweed *(Gutierrezia sarothrae)*, plains prickly pear *(Opuntia polyacantha)*, wolfberry *(Lycium cooperi)*, and perennial grasses such as indian rice grass *(Orizopsis hymentoides)*, desert needlegrass *(Stipa speciosa)*, galleta grass *(Hilaria jamesii)* and dropseed grass *(Sporobolus cryptandrus)*.

Animal species representative of the Great Basin Desert include pronghorn *(Antilocarpa americana)*, sage grouse *(Centrocercus urophasianus)*, the Great Basin gopher snake *(Pituophis melanoleucus)*, the sagebrush lizard *(Sceloporus graciosus)*, the short-horned lizard *(Phrynosoma douglassi)*, the Great Basin pocket mouse *(Perognathus parvus)*, the Chisel-toothed Kangaroo Rat *(Dipodomys microps)*, and the grey flycatcher *(Empidonax wrightii)*.

Big sagebrush is the Joshua tree of the Great Basin Desert. It is the most widespread and most abundant shrub of any desert in the Southwest. It marks Great Basin Desert territory with sheer biomass and the powerful smell of sage.

PLANTS
The Colorado Plateau Semidesert

The Colorado Plateau Semidesert consists of desert-adapted species of the Colorado Plateau. Most ecologists consider the plant and animal communities of this region's desert lowlands to be an extention of Great Basin Desert. The argument is justified in most cases. Exceptions include those isolated areas that support combinations of species that appear to be endemic or restricted to the Colorado Plateau. Desert grasslands, for example, are more diverse, abundant, and consist of different associations than exist anywhere in the Great Basin Desert. The Galleta Grass *(Hilaria jamesii)*/Threeawn Grass *(Aristida longiseta)* association in particular, is unique to the region.

In addition, there are numerous plant species that exist only on the Colorado Plateau. These include several species of locoweed *(Astralagus sp.)*, cryptantha *(Chryptantha sp.)*, and buckwheat *(Eriogonum sp.)*. Animal species endemic to the region include: the White-Tailed Prairie Dog *(Cynomys gunnisoni)*, the Mesa Verde Night Snake *(Hypsiglena torquata)*, the Midget Faded Rattlesnake *(Crotalus viridus concolor)*, the Plateau Whiptail Lizard *(Cnemidophorus velox)*, and the Painted Desert Glossy Snake *(Arizona elegans)*.

Cliffrose and Desert False Wilow.

THE PLANT KEY

The St. George Basin forms a topographic low-spot in Utah. Elevations average between 2500 and 3000 feet maintaining a growing season that lasts nearly two months longer than anywhere else in the state. This creates a hospitable climate for Mojave Desert and Great Basin Desert species. It's also one of the wettest of the Mojave habitats with an average annual rainfall of nearly eight inches. The increased rainfall combined with the transitional location between deserts, and the proximity of biota sustained by the surrounding highlands, creates one of the most diverse desert environments in North America.

IDENTIFY THE PLANTS

The following presents a key to the identification of most of the native woody plants (trees and shrubs) found in our area. Their presence generally bears the mark of biotic communities characteristic of the Mojave Desert, the Great Basin Desert, and the Colorado Plateau Semidesert. Detailed descriptions of the plants are presented in Appendix A, in alphabetical order.

To use the plant key, find a plant, read each of the following four statements, and decide which statement best describes the plant you are looking at.

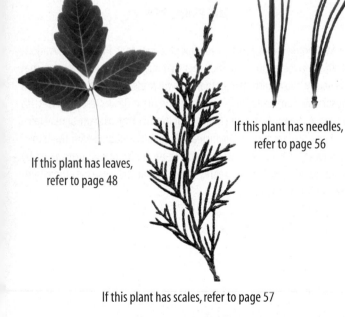

If this plant has needles, refer to page 56

If this plant has leaves, refer to page 48

If this plant has scales, refer to page 57

If this plant does not have leaves, needles, or scales (no foliage whatsoever) turn to page 57

Claret Cup

Palmers Penstemon

Brittlebrush

Mojave Prickly Pear

Desert Baylea

PLANTS

Decide what kind of leaf you are dealing with by choosing from the following categories of leaves:

Leaves with spines,
refer to page 54

Leaves (or leaflets) with lobes,
refer to page 49

Leaves (or leaflets) with jagged
edges, refer to page 50

Leaves divided into tiny leaflets,
refer to page 51

Leaves that don't fit into any of the categories listed above,
refer to page 51

LEAVES WITH LOBES

Bitter Brush

Sagebrush

Desert Baileya

Creosotebush

White Bur Sage

Virgin's Bower

Mulberry

Gambel Oak

Arizona Grape

Globemallow

Arizona Sycamore

California Gourd

LEAFLETS WITH LOBES

Squawbush

Box Elder

LEAVES WITH JAGGED EDGES

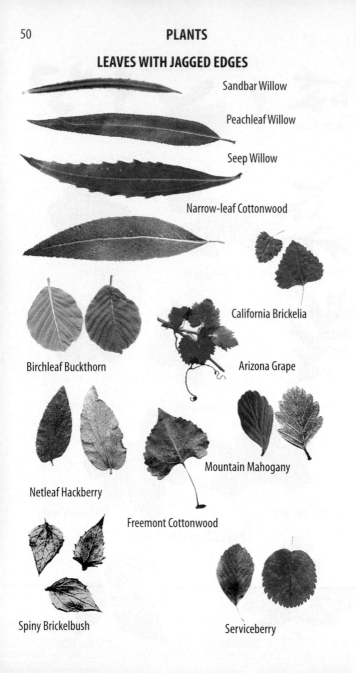

Sandbar Willow

Peachleaf Willow

Seep Willow

Narrow-leaf Cottonwood

California Brickelia

Birchleaf Buckthorn

Arizona Grape

Netleaf Hackberry

Mountain Mahogany

Freemont Cottonwood

Spiny Brickelbush

Serviceberry

LEAFLETS WITH JAGGED EDGES

Velvet Ash   Hoptree   Box Elder   Tree of Paradise

LEAVES DIVIDED INTO TINY LEAFLETS

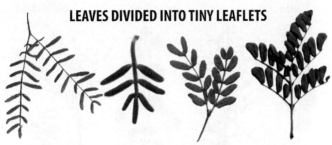

Honey Mesquite   Indigobush   Screwbean Mesquite   Catclaw

LEAVES THAT DON'T FIT INTO OTHER CATAGORIES

Leaves and branches opposite
one another Page 52

Leaves round and flat page 52

Leaves thread-like or cylindrical page 53

Others page 53

LEAVES AND BRANCHES OPPOSITE EACH OTHER

Bladdersage

Desert Sage

Buffalo Berry

Silk Tassel Bush   Buckbrush   Single Leaf Ash

Blackbrush

LEAVES ROUNDISH AND FLAT

Redbud

Shadscale

Hopsage

Brittlebrush

LEAVES NARROW-TO-THREAD-LIKE OR CYLINDRICAL

Anderson Thornbush

Burrobrush

Ratany

Iodinebush

Snakeweed

Thread-leaf Groundsel

Buckwheat

Sand Sage

OTHERS - NARROW LEAVES

Arrow weed

Russian Olive

Desert Almond

Four-wing Saltbush

Greasewood

Rabbit Brush

Desert False-

Winter Fat

Goldenbush

PLANTS

OTHERS--NOT AS NARROW

Pallid Wolfberry

Russian Olive

Sacred Datura

Desert Plume

Manzanita

Coyote Mellon

LEAVES WITH SPINES
(Oaks and Barberries)

Barberry

Oaks

AGAVE AND YUCCA
Leaves Marginally Spined and Succulent

Agave, Century Plant

Leaves Not Succulent or Marginally Spined

Leaves narrower than one inch wide

Joshua Tree Narrow-leaf Yucca Utah Yucca

Narrow-leaf and Utah Yucca differ in that the former usually has leaves less than 18 inches long. Joshua trees are usually tree-like with leaves less than 12 inches long.

Leaves an inch wide or more

Mojave Yucca **Banana Yucca**

Shigidera generally has a well defined trunk and is sometimes treelike. It only grows in the low deserts. Banana Yucca usually doesnt have a trunk and can be found at low and high elevations.

Plants With Needles (The Pines)

Pine trees can be differrentiated by the number of needles born on their fascicles:

Fascicles sheathing bundles of one, two, and three needles

Pinyon Pines
(One and Two Needle Pines)

Single-leaf pinyon

Two-leaf Colorado Pinyon

Pinyon Cone

Ponderosa Pine

Ponderosa Pine Cone

Ponderosa Pine Needles

PLANTS WITH SCALES

Utah Juniper

Tamarisk /Salt Cetar

PLANTS WITHOUT NEEDLES LEAVES OR SCALES

Ephedra Species

Joint Fir

Mormon Tea

PLANTS

The Cacti

Cacti With Tear Drop Shaped Pads

Plains Prickly Pear

Beaver Tail
Prickly Pear

Berry Prickly Pear

Barrel Shaped Cacti

Barrel Cactus   Hedgehog   Claret Cup

Other Cacti

Golden Cholla

Pin Cushion

PLANT DESCRIPTIONS

Anderson Thornbush *(Lycium andersonii)*. This medium-sized (waist-heigh give or take a foot) shrub has leaves that are generally tear-drop shaped and succulent. The thorny branchlets may look like stems at times. A few juicy orange or red edible berries may be growing on the plant. They are quite palatable, although filled with numerous seeds.

Arizona Grape *(Vitis arizonica)*. The heart-shaped leaves of this plant have jagged edges. Opposite many of the leaves is a tendril, a modified leaf that looks like a length of string that somehow manages to wrap itself around nearby objects and pull the plant upward towards the sunlight. Arizona Grape produces small purple to black grapes, about 1/2 inch wide that grow in clusters. They are edible and quite tasty but take a little getting used to.

Arizona Sycamore *(Platanus arizonica)*. This tree often grows over 70 feet high. It produces dense, golf ball-sized clusters with a spiny exterior and cottony interior. The native species grow wild along permanent desert streams and rivers in Arizona and New Mexico. The white, flaky bark, golf ball like fruit clusters, and its large size distinguish the sycamore. This tree makes an excellent shade tree. They are often planted along roads of residential areas in the desert Southwest.

Arrow Weed *(Pluchea sericea)*. The leaves on this shrub are sessile. In other words, there are no leaf stems attaching the leaves to the branches. The leaves often vary in width depending on conditions. This plant can be bushy, but generally all the branches grow vertically and straight as an arrow. Native Americans groups often used the stems as arrow shafts. You'll only find this plant near a permanent or semi-permanent water source.

Big Sagebrush *(Artemisia tridentada)*. This plant has pale-green foliage. It is the most abundant and wide-spread shrub in the Great Basin Desert. It is a member of the sunflower family but smells a lot like sage species of the mint family. It likes higher elevations and becomes the dominant life form beginning in the northern section of Snow Canyon northward into the Great Basin Desert. The plant has traditionally been used as a "smudge", by native cultures to "clean the air" of unfavorable spiritual energy. Leaf-bearing branchlets are first dried, then wrapped and burned like incense.

Birchleaf Buckthorn, Coffeberry *(Rhamnus californica)*. The leaves on this medium-to-large shrub are bright green on top, paler on the bottom, deeply veined, and appear somewhat birch-like. The youngest twigs are red or greenish and covered with tiny hairs that make it wooly to the touch. The mature branches and trunk are light to dark gray. The plant produces red berries that turn black and purple in the fall. The berries are too bitter for human consumption but provide an important food source to many bird species. As one of its common names imply, the dried berries were once used as a substitute for coffee. They also contain cascarin, a mild laxative.

Bitterbrush (Purshia tridentata). The three lobes on the leaves of this medium to large shrub are on the tips of the leaves, like sagebrush, except the leaves are wider and olive to bright green and they don't smell like sage. This shrub produces an abundance of small, bright, rose-like yellow flowers in the late spring and early summer.

Blackbrush (Coleogyne ramossissima). This small shrub (knee-height or shorter) with thorny branchlets has twigs that are opposite one another. They occur almost at right angles to the stems from which they arise. Like Ratany it may be leafless or nearly so during much of the year and may resemble a pile of sticks. The older bark is rough textured and black, while the new stems are ashy gray. When flowering, it produces an abundance of small, yellow rose-like flowers.

Bladder Sage (Salazaria mexicana). This plant is fairly distinctive. Its leaves are always opposite one another, and its fruits look like little, white paper bags. The plant varies in height, but will usually never grow higher than your waist. The leaves smell a little like lavender when crushed between the fingers.

Box elder (Acer negundo). The leaves of this tree consist of three leaflets that are lobed or serrated. It grows along streams and water courses throughout our region. It is a cousin of the maple and provides shade and produces furniture-grade lumber as well as maple syrup. The syrup is extracted in the late winter and early spring, preferably, just before the leaves begin to develop. A hole must be bored into the sapwood on the sunny side of the trunk. A spigot consisting of a wooden dowel or similar-shaped object with a groove on its upper side should be inserted into the hole, with a bucket hung below the spigot. The process is slow and laborious, and needs to be carried out on several trees, but will eventually render a substantial quantity of sweet liquid which can be boiled down into maple syrup.

Brittlebrush (Encelia farinosa). The leaves on this shrub are blue-green to gray and occasionally soft and fuzzy. The branches are quite brittle. In the dry season this plant may look like a pile of sticks, but when it rains, this small to medium-sized shrub will produce an abundance of bright yellow, daisy-like flowers. After the flowers die, the flower stalks usually remain elevated about 4 to 12 inches above the plant body.

Buckbrush *(Ceanothus montanus).* This large shrub grows at the higher elevations of our area. The leaves are roundish, smooth, leathery, greenish-gray and concave above, and grayish underneath. It produces an abundance of white, showy, sweet-scented flower clusters.

Buckwheat Bush *(Erigonum fasciculatum).* The durable and minutely hairy leaves on this small shrub emerge in clusters of three to nine leaves per leaf node. Rust-colored flowers often remain on the plant through the winter. The bark on this plant sheds in long strips.

Buffalo Berry *(Shepherdia rotundifolia).* The leaves on this medium to large shrub are round or ovalish and appear to be covered with a white powdery substance, making the plant appear grayish-green. They seem to like canyon hillsides. The plant has male and female flowers. The females produce berries that have been exploited by native cultures but are extremely sour. The plant sometimes has thorny branchlets, but it may be hard to discern because the foliage can get pretty dense.

Burrobrush *(Hymenoclea salsola).* This plant's foliage is a little flimsy looking and, generally, yellow-green or green. The margins of its leaves are folded downward. It has little, papery, white flowers, no more than an inch wide, that bloom in the spring and persist on the plant through the season. It loves disturbed ground and can be found on roadsides, empty lots, fallow fields, and ditch banks at low elevations. The crushed foliage releases a nauseous odor.

Bursage *(Ambrosia dumosa).* This low growing shrub has many, brittle branches and silvery leaves that are divided into five or six narrow lobes. During the dry season, this shrub may look like a pile of sticks. Often, you will be able to see small, green, bell-shaped flowers that become somewhat spiny with age. It's actually a species of ragweed. Where you find one of these, you will generally find hundreds, as it is often a dominant shrub in low desert landscapes.

California Brickellia and Spiny Brickellbush *(Brickellia californica, Brickellia atractyloides).* It has rough textured leaves covered with a layer of tiny hairs. The foliage is aromatic when crushed between the fingers. It produces clusters of yellowish white to greenish flowers that emerge from the leaf axils. They are sometimes tinted purple. They prefer rocky hillsides near seasonal washes. California Brickellia grows to three feet high and has heart-shaped leaves. Spiny Brickellbush has smaller, triangular-shaped leaves with jagged edges.

California Gourd *(Cucurbita palmata)*. The leaves on this shrub are rough textured, triangular and about six to ten inches long. It produces large white, tubular flowers and baseball-sized fruit that look like little round squashes. It is a relative of the squash family and can be exploited as an emergency food resource much like that described for Coyote Melon. It likes lower elevations than its cousin and can be found throughout St. George and Bloomington.

California Redbud *(Cercis occidentalis)*. In the wild, this is usually a small tree, no taller than 16 feet. Its flowers are purplish-pink and showy. It produces bright red, purplish, or sometimes even brown, stringbean-like pods. It's relatively rare in our area but can be found in sheltered canyons in Snow Canyon up to about 5,000 feet. It has become fairly popular as an ornamental. It is sometimes hard to find in the wild. Native Americans roasted the pods and ate the seeds.

Catclaw *(Acacia greggii)*. This plant has hooked, catclaw like spines. They alternate, one at a time, along a given branch. The tiny leaves are twice compound (leaflets divided into even smaller leaflets). This is one of the most annoying plants of the desert southwest because of its nasty, clothes-seeking hooks. If you plan to pass through a thicket of these, bring a bulldozer.

Century Plant *(Agave utahensis)*. This group of plants look a lot like yuccas. The leaves are rigid, sword-shaped, succulent, and are lined and tipped with sharp, woody thorns and spines. The plant flowers only about once every ten years or so. It sends up a flower stalk that often grows over 10 feet in a matter of two or three weeks. The flower stalk produces a large panicle of white flowers and seed pods, afterwhich the plant dies. The plant was an important food source for Native Americans who roasted the "hearts," i.e., the bulbous portion of the plant from which the stalks emerge. Close relatives of our local Agave are exploited in Mexico by mescaleros for making pulque, tequila, and mescal.

Cliffrose *(Cowania stansburiana)*. This medium to large shrub may sometimes resemble a tree. Its leaves are divided into five to seven lobes and are slightly whitish and hairy on one side. The leaves crowd together toward the tips of the branchlets. If you hold a leaf up to the sun, you will see small black dots that often cover the leaves. If you crush one of the leaves between your fingers, it gives off a pungent odor. When in bloom, this shrub is covered with small yellow to off-white blossoms. The bark on the older shrubs is conspicuously shreddy.

Coyote Mellon *(Cucurbita foetidisima)*. This vine has leaves up to 12 inches wide and 15 inches long. It produces large, yellow flowers and baseball-sized fruit that look like little round squashes. The plant is related to the squash family. It appears to prefer disturbed ground like roadsides and abandoned fields and is widespread in our area. The pulp stinks, as the Latin epithat foetidisima implies, and is poisonous. The seeds were reportedly roasted and eaten, however, by numerous native cultures and may be worth a try. It is also reported that the large bulbous root, sometimes referred to as "old man in the ground," is edible when cooked. The labor would be exhausting as the large tap root of the mature plant is buried several feet below the surface, but could potentially offer a substantial quantity of food to the ambitious forager.

Creosotebush *(Larrea tridentata).* The leaves on this shrub have two lobes per leaf. Observe closely. The leaves are small and the lobes may look like two separate leaves. This shrub is not heavily foliaged and may appear sticky and covered with resin. In our area, this plant is usually limited to lower elevations in the Mojave Desert, on flat terrain and well-drained soils. It is by far the most widespread and dominant plant of North America's low deserts, covering literally millions of acres in the Mojave, Sonoran, and Chihuahuan Deserts. Several of its main branches arise near or right out of the ground. It has yellow flowers during the spring and produces a strong odor. Native cultures considered this plant a panacea or cure-all. Recent research has demonstrated that this plant contains powerful antimicrobial properties.

Desert Almond *(Prunus fasciculata).* This shrub grows up to six feet tall, has small, tear-drop-shaped leaves and produces poisonous, fleshy, almond-like fruit. The plant has an abundance of thorny branchlets and looks a bit like Wolfberry, but the leaves are not succulent and it do not produce succulent red berries.

Desert Baileya *(Baileya pleniradiata).* This small plant has distinctly lobed, wooly leaves that emerge in small rosettes. Throughout the growing season, an abundance of showy, yellow, marigold-like flowers emerge on solitary stalks well above the seemingly insignificant rosettes. They are widespread in the low deserts and make great ornamentals. They are even cultivated in some western cities.

Desert False Willow *(Chilopsis linearis).* This tree grows up to 30 feet high but is usually much smaller. It produces white, purple, and showy flowers and long, dangling pods. It generally likes elevations below 3,000 feet along desert washes, streams, and rivers. This tree looks a lot like a willow, especially when there are no pods or flowers on it. Unlike true willows, the leaves have smooth margins. This tree has become fairly popular among xeriscapers for its economical use of water and showy flowers.

Desert Plume *(Stanleya pinnata).* The leaves on this plant are grayish. It erects a long flowering stalk with elongated, almost needle-like fruits. The stalk ascends well above the leaves and will sometimes be taller than you.

Desert Sage *(Salvia dorii).* This is a small, silvery shrub with four-sided branches and branchlets. It's a member of the mint family. If you crush a few of the leaves between your fingers, it will impart an almost overpowering scent of sage. Its flowers are purple and appear to arise directly from the leaf nodes. This plant thrives in small canyons and steep slopes in the hot arid climates of the Mojave Desert. Zane Gray's novel "Riders of the Purple Sage" refers to this plant.

Four-Wing Saltbush *(Atriplex canescans).* This plant is extremely widespread and often grow waist-high or higher. It has gray-green narrow leaves. On close inspection, the leaves may appear to be covered with tiny black dots. The most distinctive characteristic is its four-winged fruits that often persist on the plant year around.

Fremont Barberry *(Berberis fremontii).* The leaves on this large shrub or small tree are divided into leaflets, usually seven or so per leaf. Often mistaken for desert holly, the leaves of this plant are spiny, leathery, and appear waxy or shiny after a rain storm. It produces bright yellow flowers in the summer and has bright yellow heart wood. The berries are edible if you get them at the right time of year. In our area, this plant is found at higher elevations, usually above 4000 ft.

Fremont Cottonwood *(Populus fremontii).* This tree grows up to 90 feet tall. It produces egg-shaped capsules filled with cottony seeds. It is wide-spread in our area, growing just about anywhere there's a permanent water source or a high water table. The plant is famous for being an indicator of water for thirsty southwestern explorers. It is prized as an ornamental and lines the streets of many southwestern towns. The trunks of the older trees can become really gnarled and furrowed and attain a diameter of up to five feet. The cotton from this tree tends to irritate anyone with allergies It fills the air and covers the ground of stream and river woodlands with a blanket of cottony hairs.

Gambel Oak *(Quercus gambelii).* This tree grows up to 70 feet but is most often a small or shrubby tree forming thickets. Like all oaks, it produces acorns. The plant is abundant in most of the southwest above 5,000 feet and in sheltered or moist canyons at lower elevations. The acorns of this oak are edible but usually need to be boiled repeatedly with several changes of water. Once leached, the acorns can be dried, ground into flour, and used with other flours to make acorn bread, muffins or pancakes. This tree also makes great, furniture-grade lumber.

Globemallow *(Sphaeralcea sp).* This plant looks more like a weed than a shrub. The leaves vary greatly and can even differ on the same plant. The leaves are usually lobed but can appear jagged. It usually has orange blossoms, but sometimes they can be red or even blue, depending on the species.

Goldenbush *(Haplopappus nana).* This shrub varies greatly in size. It looks a lot like snakeweed but it's usually bigger and the foliage is concentrated on the upper half of the plant. It is very wide-spread in our area, especially at lower elevations.

Greasewood *(Sarcobatus vermiculatus).* This shrub likes saline or alkaline soils. Its presence signals the close proximity of groundwater, usually less than 20 feet. It is an intricately-branched shrub with narrow, succulent, evergreen leaves. The smaller branches eventually become sharp thorns. Early greasewood foliage was reportedly exploited local natives who boiled the leaves as a potherb to satisfy their salt requirements.

Honey or Screwbean Mesquite *(Prosopis glandulosa or Prosopis pubescens).* This plant can be a tree or a shrub. It has nasty spines paired at a node. Each leaf Is divided into several small leaflets. There are two common species of this thorny shrub in our area. These can be most easily distinguished by their pods and by the color of their bark. Long, yellow pods and dark brown bark will identify honey mesquite, while screwbean pods and light-brown bark identify Screwbean Mesquite. The pods of Honey Mesquite were a food staple of numerous indigenous cultures of the Southwest. The seeds must be removed, however. The pulp from the pods of both species is sweet and nutritious and can be used to make syrup, breads, and even an intoxicating beverage.

Hop Tree *(Ptelea angustifolia).* This plant grows up to 20 feet high. The plant is relatively rare in our area, but can be found in moist canyons and valleys up to about 5,000 feet. The leaves are a bit hairy when young and typically dark green above and paler beneath. It produces drooping clusters of disc-like papery fruits with a seed in the middle. Early settlers claimed that the fruits could be used as a substitute for hops in such enterprises as making beer.

Hopsage *(Grayia spinosa).* The new twigs on this small to medium shrub are reddish and may appear to have white strings of bark attatched to them. The leaves come in lots of different sizes but are often spoon-shaped. After the plant flowers, its fruits look like hops.

Indigobush *(Psorothamnus fremontii).* This small to medium-sized shrub has leaves that are divided into small leaflets, usually five to seven. It has sprawling, tangled, white branches and produces conspicuous, dark violet flowers in the spring.

Iodine Bush *(Suaeda toreyenna var. ramosissima).* The stems, leaves and inconspicuous flowers on this plant are all green. It grows about anywhere the soil is alkaline and/or salty and moist. Prehistorically, the seeds were occasionally exploited for food and boiled in water as a gruel, like cracked wheat. A black dye was extracted from the boiled foliage.

Joint Fir/Mormon Tea *(Ephedra trifurca and Ephedra nevadensis).* This shrub has no discernible leaves, just numerous, greenish, jointed stems, which function like leaves and photosynthesize. Early Mormon settlers boiled the jointed stems to make a stimulating tea. The plant is reported to contain an alkaloid similar to ephedrine, a stimulant used in modern pharmacopia.

Manzanita *(Arctostaphylos pungens).* This medium to large shrub has bright red bark all over it. The leaves feel smooth and waxy or leathery and will snap in two if you fold them in half. The fruit looks like a miniature apple. The miniature apple-like berries are edible if cooked and were often exploited by Native American cultures of the region in times of drought or food shortages. The dried leaves have been used as a substitute for tobacco. They reportedly impart a somewhat narcotic effect.

Mountain Mahogany *(Cercocarpus betuloides).* This plant can grow up to 20 feet high but is usually a medium-sized shrub. The flowers are yellow and usually about dime sized. The fruit consists of a narrow feather-like appendage attached to a seed. It usually grows on dry, rocky soils of mountain slopes and foothills. The leaves' jagged edges are generally confined to the margins of the top half of the leaves. This is similar to Service Berry leaves (Amelanchier utahensis). They can be distinguished by the fact that the leaves are evergreen, and dark green above and paler green beneath. The wood of this tree is very hard. Indians used the branches for digging sticks and for ax handles.

Mulberry *(Morus alba).* This tree is an introduced species, planted by pioneers and subsequent settlers in the St. George region. It was originally planted with the purpose of cultivating silk worms, which feed on the foliage, in an attempt to develop a silk industry. It grows up to 40 feet or higher and is prized as a shade tree and for its edible reddish-black fruits that resemble blackberries and raspberries. This tree can be found in most lower-elevation southwestern cities.

Narrow-leaf Cottonwood *(Populus angustifolia).* This tree grows up to 50 feet high, or sometimes more. It produces small catkins and egg-shaped capsules filled with cottony seeds. It prefers mountains streams, above 5,000 feet, or moist canyons. The leaves look a bit willowish with their minutely jagged edges. But narrow-leaf cottonwoods have a slightly blunt or rounded leaf tip, whereas willow leaves come to an acute point at the tip. The size, the deeply furrowed bark and the fruits of the older trees are unique to cottonwoods. The new leaf buds smell like balsam.

Netleaf Hackberry *(Celtis reticulata).* This tree grows up 20 feet high. It has red to orange, raisin-sized or larger edible fruits. The fruits consist mostly of pit, however, and would require consumption of large quantities if survival was an issue. The thin pulpy exterior tastes a bit like dates. It inhabits desert watercourses and drainages, and often thrives in wet canyons, up to 5,000 feet in elevation. The leaves are usually a little uneven at the base, one side of the leaf base being slightly higher or lower than the other. Also, there will often be little galls on the leaves, bumps near the base where the leaves have been infested by hackberry-loving insects.

Pallid Wolfberry *(Lycium pallidum).* This small to medium-sized shrub is extremely thorny. It has evergreen, bluish-white, waxy leaves that are succulent. It produces green-ish flowers up to an inch long that eventually develop into red, raisin-sized berries. The fruit is edible but is often bitter. The plants tend to be abundant wherever they have established a foothold. They like desert washes at low elevations where their relatively high water requirements can be met. At higher elevations, the plant can occasionally be found on rocky slopes and plains.

Pinyon *(Pinus edulis or monophylla).* This tree grows up to 35 feet but is usually much smaller. It produces woody pine cones that contain pine nuts. It occurs all over the Southwest, on rocky slopes and in the mountains, usually between 5,000 and 7,000 feet. Single-leaf pinyon is confined to the western-most part of the Southwest (Nevada and western Utah), while Colorado pinyon grows everywhere else. The nuts are tasty and com-mercially harvested. When abundant, pinyon nuts were a staple food of Native Americans.

Ponderosa Pine *(Pinus ponderosa).* This tree grows up to 130 feet high. It produces woody, fist-sized, pine cones with spines. It grows almost anywhere in the Southwest between 7,000 and 9,000 feet, especially on high plateaus. The bark of the younger trees is black which is why lumberjacks refer to them as "black jacks." On mature trees, the bark becomes orange to reddish-brown, scaly and smells like vanilla or cinnamon. This tree is the most wide spread pine tree in the Southwest. Its wood is so valuable that clear-cut-ting of entire ponderosa forests has not been uncommon. You'll notice that most pon-derosa forests are young, usually 70 years old or less. This is because most of the Southwest's ponderosa pine forests were clear-cut in the early 1900's.

Prickly Pears *(Opuntia sp.).* This group of cacti usually has tear-drop shaped pads cov-ered with spines. They produce relatively large, bulb-like fruits that can vary in color from green-to-yellow-to-red. The fruit and the pads of most species are edible but require con-siderable processing to remove the spines before they can be eaten. Local Native American groups stirred the pads and fruits in hot coals to facilitate the process. The fruits and pads of some species are sold in grocery stores. Called nopales, the pads are a food staple in Mexico and are often served with eggs for breakfast. The fruits of one local species, Berry Prickly Pear, are particularly sweet and juicy if one can tolerate the numerous large seeds.

Rabbit Brush *(Chrysothamnus nauseosus).* This plant is a disturbance species and is most common in habitat previously dominated by Big Sagebrush. It generally dominates roadsides that pass through sagebrush habitat. It produces clusters of small golden flowers. Usually, it will have shreddy bark, a strong odor, and twigs covered with tiny hairs. This plant will often be mistaken for snakeweed, although it is a great deal larger.

Ratany *(Krameria parvifolia).* This small shrub is leafless during much of the year and may resemble a pile of sticks. It's thorny branchlets are slender and sharp. It is extremely common in some areas of the Mojave Desert, often growing with Creosotebush and Anderson Thornbush. It blooms red or lavender during flowering season and produces a roundish, almost egg-shaped pod covered with redish prickles that may persist on the plant long after the leaves and flowers fall off.

Russian Olive *(Elaeagnus angustifolia).* This tree grows up to 20 feet high in disturbed ground along watercourses and roadsides. It has silvery leaves that are darker on top and paler underneath. The bark on the younger twigs is often reddish and is usually armed with thorns. The fruits resemble small olives and are edible but tasteless and consist mostly of a large pit with a thin, pulpy exterior. The plant is native to southern Europe and western Asia, but thrives in native southwestern communities. It often out-competes native flora, taking over the territory of native trees and shrubs.

Sacred Datura *(Datura metuloides).* The leaves on this small to medium-sized shrub are large, up to twelve inches long and velvety. This is an extremely conspicuous plant when in bloom. It will bear large, 4 to 8-inch-long, white, trumpet-shaped flowers that will eventually become thorny, golfball-sized burrs. Danger! Every part of this plant is extremely poisonous. The plant contains an hallucinogenic alkaloid that can cause visions and sometimes death.

Sand Sagebrush *(Artemisia filifolia).* This shrub has gray-green foliage. Seen from a distance it appears almost blue. It likes hot, arid conditions and deep sand, and can out-compete almost any other life form in this meager habitat for the limited resources available. Its needle-like leaves are flimsy and make the plant appear droopy. Look for this plant in sand dunes. As one sniff of the crushed leaves will suggest, this is a close relative of Big Sagebrush.

Seep Willow *(Baccharis glutinosa).* This medium to large shrub usually grows near streams, rivers and ephemeral washes. If you turn one of the leaves upside down, you will notice that it has three conspicuous veins that run from the base to the tip of the leaf. Sometimes you need to look at several leaves before you can really see the veins. Usually, most of the leaves will be toothed, especially the older ones. The number of teeth on the margins of the leaves will generally vary from two or three to several. Sometimes the leaves are sticky and a little resinous.

Serviceberry *(Amelanchier utahensis)*. This plant grows to 15 feet high, but most often looks like a shrub. It has tiny, white to pinkish, rose-like, showy flowers. It produces marble-sized, sort of apple-shaped, purple or black fruits. Usually, it grows in semi-moist soils above 4,000 feet. It is often abundant in mountains in forest openings where shrubs, rather than juniper and pinyon pine dominate. This plant is most easily identified because the leaves' jagged edges are confined to the margins of the upper half of the leaf. The berries are edible and make a decent survival food

Shadscale *(Atriplex confertifolia)*. This extremely widespread thorny shrub is generally small, less than the height of your knee. It has gray-green foliage and straw-colored stems. The leaves are ovalish to round, relatively rigid and sometimes covered with small scales (look closely). It produces dime-sized or smaller, two-winged fruits. The leaves may taste a bit salty.

Silk Tassel Bush *(Garrya flavescens)*. The leaves on this shrub are arranged opposite one another. they may feel silky and hairy, but always leathery and often waxy in texture and will snap if you fold one. It is distinctively gray-green. It bears long, drooping, silky tassels after it flowers. It prefers relatively moist canyons and hillsides and generally occurs in our area above 5500 feet

Single-leaf Ash *(Fraxinus anomala)*. This plant grows up to 25 feet high but is usually much smaller. The leaves and branches are always opposite one another and the leaves on the same plant often vary in shape and size. This plant sticks out in dry desert landscapes because it is often the only tree around and the leaves are bigger than those of most desert shrubs. They are especially noticeable in the fall when the leaves turn bright yellow. It likes rocky slopes, hillsides, and dry canyons of the upper desert environments. It's common in some pinyon-juniper woodlands to 6,500 feet in elevation, but sometimes grows higher. It produces drooping clusters of key-like fruits called samaras that consist of seeds attached to papery wings that aid in wind dispersal.

Snake Weed *(Gutierrezia sarothrae)*. This small, yellow-green to green shrub is extremely common and seems to take over a plant community, especially where cows have been grazing. It usually has an abundance of upright, slender branches. The leaves are sometimes shiny and sticky.   It produces numerous small, bright yellow flowers during the spring. After the flowers die, the flower stalks usually remain elevated above the plant body.

Squaw Bush *(Rhus trilobata)*. Each leaf will generally be divided into three leaflets and will be somewhat shiny on top and dull on the bottom. This is a very bushy shrub. It bears clusters of little red berries. They are edible but rather tart-tasting and can be used to make a lemonade-type beverage.

Tamarisk or Salt Cedar *(Tamarix pentandra)*. This tree is almost always less than 20 feet high. It produces numerous pink, showy flowers. It grows along streams, washes, rivers, and moist, disturbed ground. It was initially brought over from from the Middle East about 80 years ago by the United States Department of Interior to control erosion. Since then, tamarisk has spread and infested most of the Southwest's desert rivers and streams, choking out native willows and cottonwoods in its expansion. It's sometimes referred to as saltcedar because of its ability to thrive in salty soils.

Thread-leaf Groundsel *(Senecio spartiodes)*. This plant looks a lot like Snake Weed. The leaves are narrow and often over three inches long. It produces a yellow canopy of flowers at the apex of the stems. It grows in sandy, dry, well-drained soils.

Tree of paradise (*Ailanthus alitssima*). This tree grows up to 80 feet high but is usually a lot smaller. It's a weedy tree. It can grow just about anywhere the ground is moist and has been disturbed. This tree has up to 24 leaflets with little notches on the base of the leaflets. This tree is not native to the Southwest, but is included here because it's so common in undeveloped areas.

Utah Juniper *(Juniperus Utahensis)*. This tree grows up to 40 feet high, but is usually less than 20 feet. It produces pale blue, marble-sized, fleshy berries with one or two seeds. It is widespread in the Southwest.usually between 5,000 to 7,500 feet, but often higher. This tree can get really twisted and gnarly looking, often appearing to be growing right out of solid rock. Most of the time, however, they look like a relatively short evergreen in nearly pure stands at lower elevations, and with pinyon at higher elevations. Some trees are as old as 2,000 years and have immense trunks.

Velvet Ash *(Fraxinus velutina)*. This plant grows up to 40 feet high along streams, wet canyons, sometimes along seasonal washes, and usually wherever there's some form of permanent water. The leaves can vary in shape but usually have just five leaflets with minutely-serrated margins. It produces drooping clusters of key-like fruits called samaras that consists of a seeds attached to a papery wings that aid in wind dispersal. The plant makes a great shade tree and can be harvested for its high-grade lumber.

Virgin's Bower *(Clematis ligusticifolia).* This plant is most recognizable in the autumn when it produces fluffy white flowers with needle-like sepals that form feathery clusters. The plant is a climbing vine, using its leaves or tendrils (modified leaves) to wrap around any object presenting an opportunity to climb upward up to twenty, thirty feet or more.

Wavy-Leaf Oak, Scrub Oak *(Quercus turbellina, Quercus undulata).* These oaks are evergreen and the leaves feel a bit leathery and sometimes appear shiny or waxy. Usually the spines are fairly obvious, but sometimes the spines are not there though the basic shape of the leaf remains the same. You will usually find acorns either on the plant or on the ground beneath the plant. The plant is widespread in our area and can be found in sheltered canyons at low elevations and mountain slopes and hillsides at higher elevations.

Willow *(Salix sp.).* There are numerous species of willow in the Southwest. They can be found along permanent water courses or springs. Most willow species have minutely jagged leaf margins. Sandbar willow often has smooth and jagged-edged leaves on the same plant. The new branches of all willows are limber and yellowish. You may confuse willow with narrow-leaf cottonwood because of the leaf shape and minutely jagged edges. The older twigs of the latter are gray and the leaf tips don't come to a sharp point like those of willow. Also, willows don't grow as tall as narrow-leaf cottonwoods. Willows contain salicilic acid, the active ingredient in aspirin, on the inner side of their bark.

Winterfat *(Ceratoides Lanata).* This plant is easily recognized because its foliage is wooly. It's called winter-fat because it provides winter forage for livestock. It's generally a small shrub, knee-high or smaller.

Yucca, Spanish Bayonet *(Yucca sp.).* This group of plants generally all have sword-shaped leaves of various widths and spiney tips. The margins of the leaves are sometimes lined with fraying fibers. Yuccas produce large, white, trumpet-shaped edible flowers and large, pod-like fruits. The fruits of bananna yucca are edible, if harvested before the seeds ripen. The should be boiled in two changes of water, the first for 20 minutes, the second for five minutes. Yuccas all contain saponins, chemicals that produce suds when mixed with water. Native Americans used the bulbous roots of the narrow-leaved yuccas to make soap and shampoo. All yuccas produce a long, flowering stalk that emerges well above the plant, often over six feet high. The Spaniards thought it resembled a bayonet, hence the name.

APPENDIX B
UNDERSTANDING ROCKS

There are basically three kinds of rocks in the world: sedimentary, igneous, and metamorphic. Igneous rocks come from volcanoes and consist of melted rock or magma belched out of the earth's mantle. The magma can ooze out of a volcano in the form of basalt, rhyolite or andesite (different types of quickly cooling lava), or it can get stuck under ground and never quite reach the surface like granite, diorite or gabbro (different kind of slow cooling lava). The Pine Valley Mountains and the cinder cones along State Highway 18 in the vicinity of Veyo and Diamond Valley were formed by volcanic activity. The latter deposited a protective layer of lava rock over much of our area's sedimentary deposits, aiding in the preservation and differential erosion of the layer-cake sedimentary deposits characteristic of our area. Underground water channels are sometimes created within zones of volcanic and/or tectonic activity. Hydrothermally heated, mineral rich waters passing through these channels occasionally deposit veins or pockets of valuable minerals such as uranium, gold or platinum.

Sedimentary rock is derived from combining tiny particles into one large rock. For example, take a mountain of igneous (volcanic) rock, pulverize it through millions of years of wind and water action, gravity, plant growth, and acid rain, and you get decomposition of the large mountain into tiny particles. From largest to smallest, these particles include boulders, cobbles, pebbles, sand, silt, and clay. The particles are then sorted by wind, wave action, or current so that they become concentrated. Under the right conditions such particle concentrations eventually become glued or cemented together into sedimentary rock. The process is called lithification and is the result of compaction from the enormous weight of superimposed formations combined with the downward seepage of groundwater, which precipitates calcium carbonate or lime, accelerating particle binding (i.e., lithification). Lithified sand particles make sandstone. Lithified silt particles make siltstone. Lithified clay particles make shale or mudstone. Add a little of everything together and you get a conglomerate.

Limestone is a type of sedimentary rock formed at the bottom of oceans where calcium carbonate minerals are separated from seawater through microbial activity and/or other chemical processes and gradually allowed to settle. Limestone can also form at shoreline from continuous buildup of marine organisms with calcium carbonate shells (like a big conglomerate of sea shells).

Most of the Dixie Basin and surrounding mountains are composed of sedimentary stone laid down during the age of dinosaurs (250 to 60 million years ago) in orderly succession like pancakes, revealing millions of years of the earth's history. Sometimes stones such as jasper, cherts, or chalcedony can form within layers of limestone, originally being deposited as a silicious, gelatinous mass that eventually forms a dense nodule.

Metamorphic rocks form when a mineral is subjected to intense heat and is compressed so hard by gravity or tectonic pressures that it literally changes into a different mineral. For example, squashed or metamorphosed sandstone makes quartzite. Metamorphosed granite makes gneiss. Metamorphosed shale makes slate or schist. The Beaver Dam mountains, just east of St. George, are metamorphic. The mountain range has made fortunes for miners exploiting its mineral resources.

LOCAL MINERALS

While the scope of this booklet does not allow for an adequate description of all metamorphic, igneous, and sedimentary resources found in our area, the following table provides an overview of valuable minerals found and exploited by miners and rock-hounds. Most of the information in the table came from *Geological Resources of Washington County, Utah* by Miriam Burgden (Public Information Series 20, Utah Geological Survey, 1993).

APPENDIX B

COMMON LOCAL SEDIMENTARY ROCKS

Conglomerate   Conglomerate

Shale   Sandstone

COMMON LOCAL VOLCANIC ROCKS

Granite   Basalt

Minerals Located in the Surrounding Area

Rock Hounding	High Tech	Precious (semi)	Other
Beaver Dam Wash and the Beaver Dam Mountains			
Malachite	Germanium	Silver	Antimony
Azurite	Uranium	Gold	Lead
Petrified wood	Tungsten	Zinc	
Agate	Gallium		
Wonder-stone			
Leeds/Silver Reef			
Agate	Vanadium	Gold	Gypsum
Petrified wood	Uranium	Silver	Lead
Bull Valley and the Bull Valley Mountains			
	Tungsten	Zinc	Iron
	Molybdenum	Copper	Arsenic
		Silver	Antimony
		Gold	
Gunlock			
	Strontium		Antimony
	Manganese		Bismuth
Santa Clara			
	Strontium	Silver	
	Manganese		
Vermillion Cliffs			
	Uranium	Gold	
		Platinum	
Bloomington Hills			
	Uranium	Copper	
		Zinc	

About the Author

Eric Hansen has spent most of his life in the canyon country of southern Utah and northern Arizona. He studied Archaeology and Botany at the University of Northern Arizona, where he received a Masters Degree in 1994. He currently runs a cultural resources consulting company (Southwest Archaeological Consultants), and works as an information systems specialist at Green Valley Spa.